Contents

How to use this book

Each page has a title telling you what it is about.

Instructions look like this. Always read these carefully before starting.

This shows you how to set out your work. The first question is done for you.

This shows that the activity is an **Explore**. Work with a friend.

This means you decide how to set out your work.

Ask your teacher if you need to do these.

Sometimes there is a **Hint** to help you.

Use doubling to help you.

Read these word problems very carefully. Decide how you will work out the answers.

Sometimes you need materials to help you.

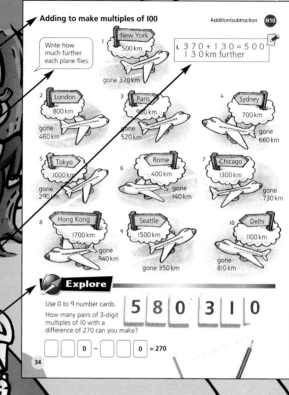

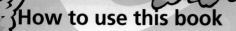

Rounding

Write the number of each team's fans to the nearest 10, 100 and 1000.

1
Ice Cats
4762

1a. 4 7 6 0 fans
b. 4 8 0 0 fans
c. 5 0 0 0 fans

2
Flyers
3954

3
Sabres
8267

4
Pirates
2318

5
Stars
6059

6
Flames
4596

7
Blades
6479

8
Phantoms
3987

9
Cyclones
9146

10
Griffins
7328

⒠ Write how many more fans are needed for each team to have 10 000 fans.

Write the total number of people at these matches.

Round each to the nearest 1000.

11.
```
    4 7 6 2
 +  8 2 6 7
  1 3 0 2 9   →  1 3 , 0 0 0
    1 1            people
```

11) Ice Cats (v) Sabres (12) Flames (v) Cyclones

13) Phantoms (v) Griffins (14) Stars (v) Pirates (15) Flyers (v) Phantoms

16) Blades (v) Sabres (17) Ice Cats (v) Flames (18) Sabres (v) Griffins

⒠ Round each total to the nearest 500.

Rounding

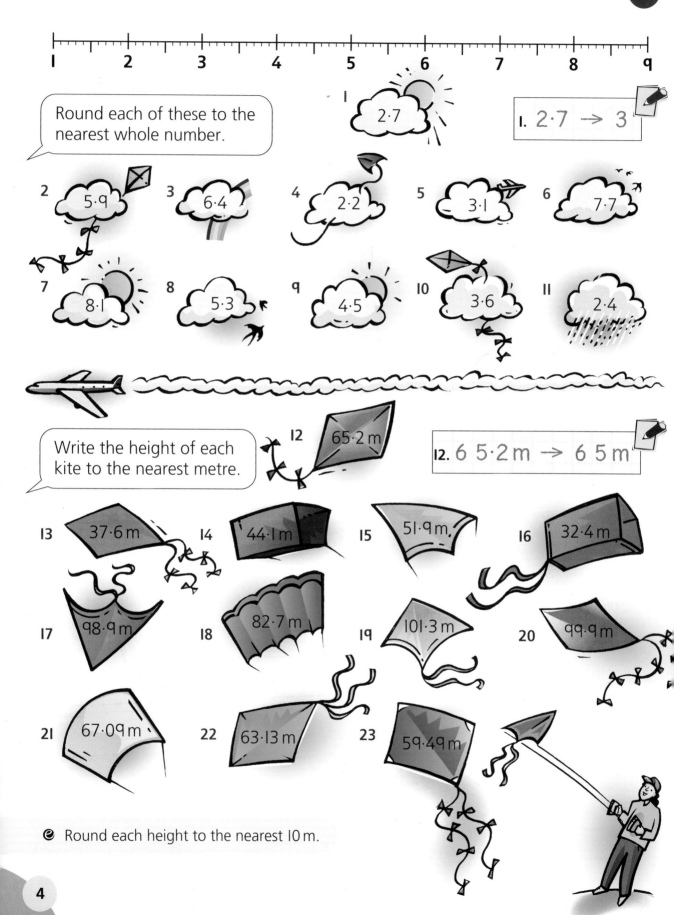

Round each of these to the nearest whole number.

1 2·7

1. 2·7 → 3

2 5·9
3 6·4
4 2·2
5 3·1
6 7·7
7 8·1
8 5·3
9 4·5
10 3·6
11 2·4

Write the height of each kite to the nearest metre.

12 65·2 m

12. 6 5·2 m → 6 5 m

13 37·6 m
14 44·1 m
15 51·9 m
16 32·4 m
17 98·9 m
18 82·7 m
19 101·3 m
20 99·9 m
21 67·09 m
22 63·13 m
23 59·49 m

ℯ Round each height to the nearest 10 m.

Adding decimals

Write how much petrol is needed to fill each tank.

1

15·7 l
18·4 l tank

1. 15·7   18·4
   0·3 ⟍ 16·0 ⟋ 2·4
   2·7 l needed

2

12·9 l
19·5 l tank

3

14·8 l
17·3 l tank

4

11·6 l
15·2 l tank

5

13·1 l
18·0 l tank

6

16·4 l
20·5 l tank

7

19·7 l
22·3 l tank

ℓ Write how much petrol is needed in millilitres and in centilitres.

Problems

8 A ball of string is **10 m** long. Dad uses **540 cm** to make a washing line. Mum uses **161 cm** to wrap a parcel. How much string is left?

9 A tree is **3·5 m** tall. It grows **50 cm** each month. How long does it take to grow to be **10 m** tall?

10 A puppy weighs **4·62 kg**. How many grams must it gain to weigh **5 kg**?

11 A marathon runner has run **17·5 miles**. A marathon is approximately **26¼ miles**. How much further must she go to complete the race?

12 Simon's train fare home is **£6**. He only has **£4·63**. How much must he borrow from his friend?

13 Anya buys a hamburger that costs **£2·87** and a drink that costs **90p**. She pays with a **£5** note. How much change does she get?

35

Adding

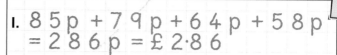

Write how much each child has spent.

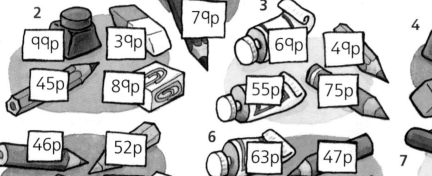

1. 58p
64p
85p
79p

I. 8 5 p + 7 9 p + 6 4 p + 5 8 p
= 2 8 6 p = £ 2·8 6

2 99p 39p 45p 89p

3 69p 49p 55p 75p

4 99p 75p 27p 61p

5 46p 52p 87p 91p

6 63p 47p 39p 90p

7 18p 78p 59p 96p

8 36p 49p 71p 68p

9 52p 77p 33p 28p

10 11p 74p 99p 47p

🅔 Write how much change each child gets from £5.

Each person pays for a weekend break and travel insurance.

II. £ 1 4 5 − £ 9 9 = £ 4 6
£ 4 6 − £ 2·9 9 = £ 4 3·0 1

The weekend break costs £99. Write how much money each person has left.

11

savings £145
insurance £2·99

12

savings £204
insurance £4·99

13

savings £149
insurance £5·99

14

savings £250
insurance £9·99

15

savings £113
insurance £3·59

16

savings £220
insurance £3·33

Adding to make 10 and 100

Each ticket is bought with a £10 note.

Write how much change there is for each.

I. £9·36 + £0·64 = £10
64 p change

TICKETS

1  £9·36

2  £9·18

3  £7·79

4  £5·54

5  £6·38

6  £7·62

7  £8·81

8  £4·01

9  £5·27

10  £3·49

11 £6·91

Copy and complete. | Write similar facts to 1000, 500, 50, 10 and 5.

12. 42 + 58 = 100
420 + 580 = 1000
420 + 80 = 500
42 + 8 = 50
4·2 + 5·8 = 10
4·2 + 0·8 = 5

12 42 +  = 100

13 73 + ☐ = 100

14 81 + ☐ = 100

15 46 + ☐ = 100

16 34 + ☐ = 100

17 27 + ☐ = 100

18 58 + ☐ = 100

19 85 + ☐ = 100

20 19 + ☐ = 100

21 93 + ☐ = 100

Adding to make multiples of 100

Write how much further each plane flies.

1 New York
500 km
gone 370 km

1. 370 + 130 = 500
130 km further

2 London
800 km
gone 460 km

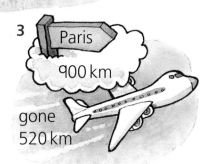

3 Paris
900 km
gone 520 km

4 Sydney
700 km
gone 660 km

5 Tokyo
1000 km
gone 290 km

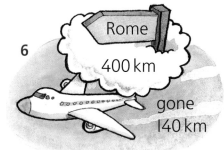

6 Rome
400 km
gone 140 km

7 Chicago
1300 km
gone 730 km

8 Hong Kong
1700 km
gone 940 km

9 Seattle
1500 km
gone 350 km

10 Delhi
1100 km
gone 810 km

Explore

Use 0 to 9 number cards.

How many pairs of 3-digit multiples of 10 with a difference of 270 can you make?

5 8 0   3 1 0

 0 −   0 = 270

580 − 310 = 270

Write the position of each hanger.

Round each number to the nearest tenth and to the nearest whole number.

a. $|\cdot||$ → $|\cdot|$
$|\cdot||$ → $|$

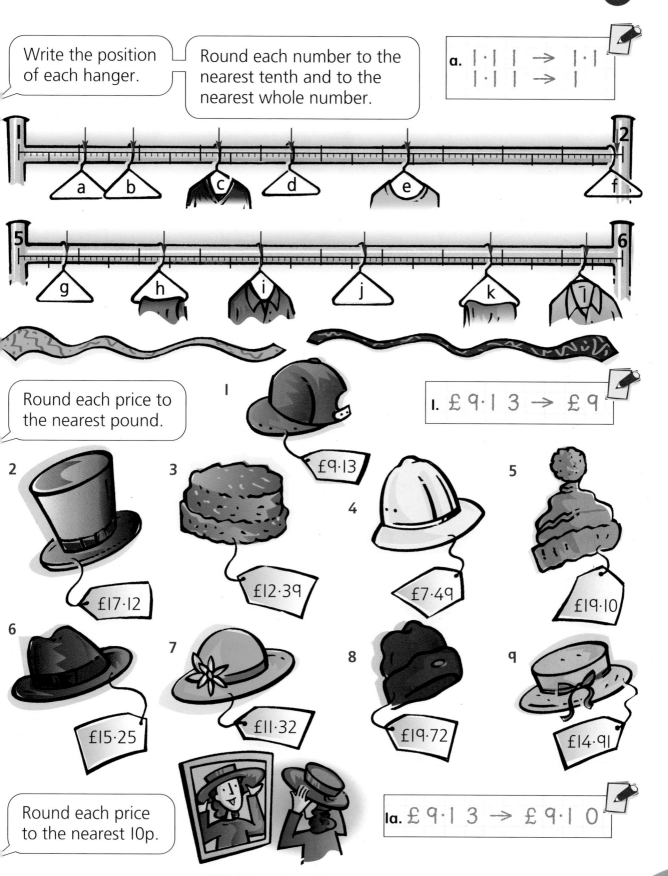

Round each price to the nearest pound.

I. £9·13 → £9

1 £9·13

2 £17·12

3 £12·39

4 £7·49

5 £19·10

6 £15·25

7 £11·32

8 £19·72

9 £14·91

Round each price to the nearest 10p.

1a. £9·13 → £9·10

e Write how much change you would get if you paid for each hat with a £20 note.

Rounding

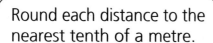

Round each distance to the nearest tenth of a metre.

1

1. $2 \cdot 2 \, 4 \, m \rightarrow 2 \cdot 2 \, m$

2·24 m

2

1·79 m

3

2·15 m

4

1·82 m

5

3·23 m

6

3·19 m

7

3·04 m

8

4·71 m

9

2·98 m

10

2·95 m

Round each distance to the nearest metre.

1a. $2 \cdot 2 \, 4 \, m \rightarrow 2 \, m$

Explore

Use number cards 0 to 9.

Make numbers with one decimal place.

☐ . ☐

How many numbers like this round to 5?

How many numbers with two decimal places round to 5?

Dividing by 10

Write the length of each insect in centimetres.

1
31 mm

I. 3·1 cm

2
22 mm

3
7 mm

4
13 mm

5
29 mm

6
4 mm

7
28 mm

8
127 mm

9
60 mm

10
48 mm

Copy and complete.

11 $38 \div 10 =$

II. $38 \div 10 = 3·8$

12 $270 \div 10 =$

13  $\times 10 = 406$

14 $\div 10 = 0·4$

15 $29 \div 10 =$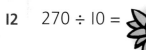

16 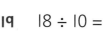 $\div 10 = 11·8$

17 $3562 \div 10 =$

18  $\times 10 = 147$

19 $18 \div 10 =$

20  $\times 10 = 9$

📧 Divide each answer by 10 again.

Dividing by 100

Write each amount in litres.

1
248 cl

I. $248 \div 100 = 2 \cdot 48\,l$

100 cl = 1 l

2
144 cl

3
399 cl

4
707 cl

5
818 cl

6
220 cl

7
401 cl

8
360 cl

9
611 cl

e Write how many millilitres in each container.

Copy and complete.

10   $341 \div 100 =$

10. $341 \div 100 = 3 \cdot 41$

11   $286 \div 100 =$

12    $\div 100 = 0.75$

13   $11 \div 100 =$

14    $\times 100 = 403$

15   $3510 \div 100 =$

16   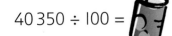 $\div 100 = 0 \cdot 09$

17   $40\,350 \div 100 =$

18    $\times 100 = 60$

19   $3 \div 100 =$

Dividing by 10 and 100

Write how many notes are in each pile.

1

£460

I. $£460 ÷ £10 = 46$

2

£6500

3

£8600

4

£9000

5

£4600

6

£4600

7

£660

Write the missing numbers.

8   $156 ÷ \blacksquare = 15·6$

8. $156 ÷ 10 = 15·6$

9   $\blacksquare ÷ 100 = 1·48$

10   $246 ÷ \blacksquare = 2·46$

11   $391 ÷ \blacksquare = 3·91$

12   $\blacksquare ÷ 10 = 1·48$

13   $\blacksquare ÷ 100 = 5·07$

14   $786 ÷ \blacksquare = 78·6$

15   $841 ÷ \blacksquare = 8·41$

16   $\blacksquare ÷ 100 = 2·11$

17   $929 ÷ \blacksquare = 92·9$

18   $735 ÷ \blacksquare = 7·35$

19   $\blacksquare ÷ 10 = 15·5$

20   $232 ÷ \blacksquare = 2·32$

Explore

Write down a 3-digit number.

Divide it by 100.

Use your calculator to divide the 3-digit number by 99.

Compare the answers. What do you notice? Does this always happen?

Try some different 3-digit numbers. Can you see a pattern?

9

Multiplying

Write how many items in each set.

I
6 eggs in a box
3 boxes

I. $3 \times 6 = 18$ eggs

2

7 pencils in a box
6 boxes

3

6 lollies in a box
9 boxes

4

8 candles in a box
9 boxes

5

5 balloons in a pack
8 packs

6

7 cakes in a tin
3 tins

7

8 crackers in a box
4 boxes

8

4 pens in a pack
7 packs

9

6 stickers in a pack
8 packs

10

9 bars in a pack
6 packs

Write the answer. Add the digits.

II. $5 \times 9 = 45$
$4 + 5 = 9$

11 $5 \times 9 =$

12 $3 \times 9 =$

13 $8 \times 9 =$

14 $4 \times 9 =$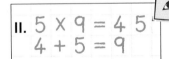

15 $7 \times 9 =$

16 $6 \times 9 =$

17 $2 \times 9 =$

18 $9 \times 9 =$

ⓔ Use doubling to multiply each number by 18. Add the digits of the answers. What do you notice?

Multiplying and dividing

Write the missing numbers.

I. $9 \times 5 = 45$
   $45 \div 5 = 9$

1 $\square \times 5 = 45$ 2 $\square \times 7 = 21$ 3 $\square \times 8 = 16$ 4 $\square \times 6 = 30$

   $45 \div 5 = \square$ $21 \div 7 = \square$ $16 \div 8 = \square$ $30 \div 6 = \square$

5 $4 \times 4 = \square$ 6 $\square \times 3 = 24$ 7 $\square \times 4 = 28$ 8 $\square \times 6 = 54$

   $\square \div 4 = 4$ $24 \div 3 = \square$ $28 \div 4 = \square$ $54 \div 6 = \square$

9 $\square \times 8 = 64$ 10 $\square \times 6 = 36$ 11 $\square \times 8 = 32$ 12 $\square \times 7 = 56$

   $64 \div 8 = \square$ $36 \div 6 = \square$ $32 \div 8 = \square$ $56 \div 7 = \square$

13 $\square \times 4 = 36$ 14 $\square \times 6 = 42$ 15 $7 \times 7 = \square$ 16 $\square \times 6 = 48$

    $36 \div 4 = \square$ $42 \div 6 = \square$ $\square \div 7 = 7$ $48 \div 6 = \square$

Write how many packs for each.

17. $18 \div 6 = 3$

17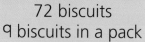
18 bars
6 bars in a pack

18
54 pencils
9 pencils in a pack

19
72 biscuits
9 biscuits in a pack

20
48 balloons
8 balloons in a pack

21
49 stickers
7 stickers in a pack

22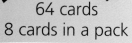
64 cards
8 cards in a pack

23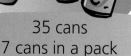
35 cans
7 cans in a pack

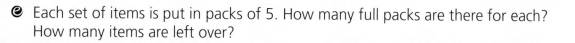

e Each set of items is put in packs of 5. How many full packs are there for each?
   How many items are left over?

Which group of children each get more?

I
| 20 nuts 2 girls | 25 nuts 5 boys |

I. $20 \div 2 = 10$
$25 \div 5 = 5$
$10 > 5$
The girls get more.

2

| 27 pens 3 girls | 40 pens 4 boys |

3
| 56 beads 8 girls | 60 beads 6 boys |

4

| 42 balloons 6 girls | 24 balloons 8 boys |

5
| 64 crayons 8 girls | 49 crayons 7 boys |

6

| 36 stickers 6 girls | 70 stickers 7 boys |

7
| 48 sweets 6 girls | 6 sweets 1 boy |

8

| 12 pencils 2 girls | 35 pencils 7 boys |

9
| 63 crisps 9 girls | 28 crisps 4 boys |

10
| 35 bars 5 girls | 54 bars 9 boys |

Copy and complete.

II $30 \div 5 =$

II. $30 \div 5 = 6$

12 $42 \div 7 =$

13 $36 \div 6 =$

14 $70 \div 10 =$

15 $48 \div 8 =$

16 $32 \div 4 =$

17 $63 \div 9 =$

18 $18 \div 2 =$

19 $56 \div 7 =$

20 $35 \div 5 =$

Explore

6 divides into 3**6** exactly.

2 divides into **2**4 exactly.

Find other pairs like this, where the first number appears as a digit in the second number.

How many can you find where the second number is less than 50?

Multiplying and dividing

Write the missing numbers.

1   $3 \times = 24$

1. $3 \times 8 = 24$

2   $6 \times = 36$

3   $4 \times = 32$

4   $7 \times = 42$

5   $ \times 9 = 54$

6   $ \times 8 = 48$

7   $ \times 7 = 56$

8   $ \times 7 = 49$

9   $8 \times = 72$

10   $ \times 4 = 28$

Problems

11 Georgie cycles to work and back **6** times in a week.

She cycles **72** miles altogether.

How far is it from Georgie's house to work?

12 Callum has **42** photos.

He can fit **6** photos on each page of his photo album. How many pages does he use?

The album has **12** pages. How many more photos can Callum stick in it?

13 Sam is making a tape. The tape is **45** minutes long and each song he records is **3** minutes long.

How many songs can he fit on the tape?

How many more would fit if he used a **60**-minute tape?

14 Yasmin saves **20p** a day.

She wants to buy a book that costs **£8·40**.

How many weeks until she has saved enough money?

15 A packet of cakes weighs **248 g**.

There are **8** cakes in a packet. How much does each cake weigh?

How many cakes would weigh just over **1 kg**?

16 A fan makes one turn every **3** seconds.

How many times does it turn in $1\frac{1}{2}$ minutes?

Doubling and halving

Shooting a basket is worth 2 points.

Write how many baskets each team has shot.

1. $\frac{1}{2}$ of $8\,4 = 4\,2$

Sharks: $4\,2$ baskets

1
 Sharks
84 points
v
 Rockets
72 points

2
Hawks
58 points
v
 Hornets
48 points

3
Bulls
56 points
v
Warriors
98 points

4
Wolves
118 points
v
 Royals
76 points

5
Kings
104 points
v
Blazers
94 points

6
Pistons
126 points
v
 Mavericks
68 points

℮ Each team shoots 11 more baskets. Write how many points they have now.

Each player shot a different number of baskets in the season.

Write how many points each scored.

7. double $4\,6 = 8\,0 + 1\,2$
$= 9\,2$ points

7
Dead-eye Dave
 46

8
Top-shot Tess
72

9
Hawk-eye Helen
39

10
Magic Mike
 57

11
Handy Andy
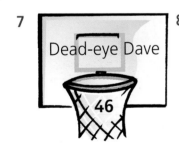 69

12
Careful Connie
 77

13
Speedy Seb
 28

14
Lightning Lil
 86

℮ Last season each player scored 26 points less. How many baskets did each shoot?

The televisions are being sold at half-price.

Write the new prices.

1. half of £320 = £160

1 £320

2  £640

3  £560

4  £180

5  £760

6  £840

7 £430

8 £670

9 £980

e As a special offer the shop gives another 50% off to the fiftieth customer. Write how much the fiftieth customer would pay for each television.

Each car was bought at half-price.

Write the original prices.

10. double £2600 = £5200

10  £2600

11  £4300

12  £9400

13 £3200

14  £1700

15  £4500

16  £6700

17  £8300

18  £3900

Doubling and halving

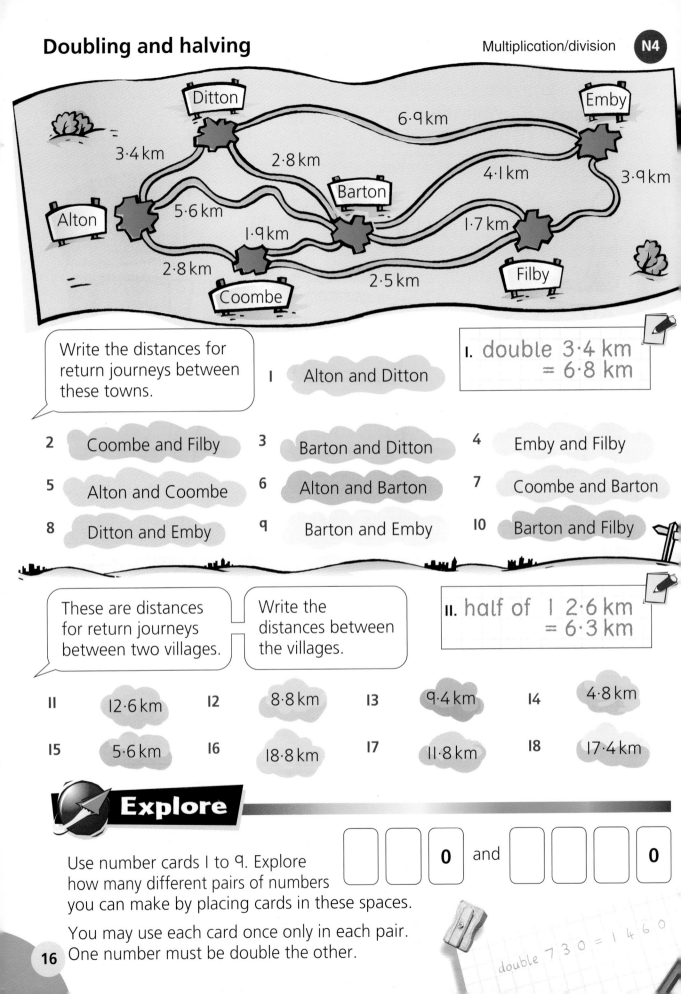

Write the distances for return journeys between these towns.

1	Alton and Ditton

I. double 3·4 km
= 6·8 km

2 Coombe and Filby	3 Barton and Ditton	4 Emby and Filby
5 Alton and Coombe	6 Alton and Barton	7 Coombe and Barton
8 Ditton and Emby	9 Barton and Emby	10 Barton and Filby

These are distances for return journeys between two villages.

Write the distances between the villages.

II. half of 1 2·6 km
= 6·3 km

11	12·6 km	12	8·8 km	13	9·4 km	14	4·8 km
15	5·6 km	16	18·8 km	17	11·8 km	18	17·4 km

Explore

Use number cards 1 to 9. Explore how many different pairs of numbers you can make by placing cards in these spaces.

☐ ☐ **0** and ☐ ☐ ☐ **0**

You may use each card once only in each pair.
One number must be double the other.

double 730 = 1460

16

> 49 children are going on a trip.

> Write the total cost for each trip.

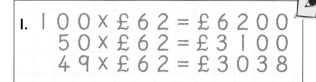

I. $100 \times £62 = £6200$
$50 \times £62 = £3100$
$49 \times £62 = £3038$

1
London
£62

2
Edinburgh
£83

3
Oxford
£24

4
Scottish Highlands
£49

5
Cambridge
£42

6
Scarborough
£53

7
Stonehenge
£27

8
Stratford
£35

9
Brighton
£42

10
Blackpool
£29

11
Lake District
£64

12
Snowdon
£57

ℯ Write the cost of 98 children going on each trip.

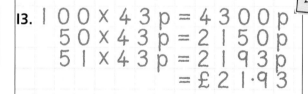

> The school is providing a snack for 51 people.

> Write the total cost for each type of snack.

13. $100 \times 43p = 4300p$
$50 \times 43p = 2150p$
$51 \times 43p = 2193p$
$= £21·93$

13
Cheese roll
43p

14
Ham salad
62p

15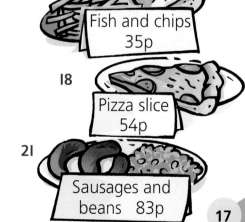
Fish and chips
35p

16
Lasagne
47p

17
Tuna sandwich
28p

18
Pizza slice
54p

19
Cherry pie
37p

20
Spaghetti
74p

21
Sausages and beans   83p

Multiplying by nearly 100

The concert organisers must decide how to arrange the chairs.

Write which arrangement has more seats.

```
1.  1 0 0 X 3 8 = 3 8 0 0
      9 9 X 3 8 = 3 7 6 2

    1 0 0 X 3 6 = 3 6 0 0
    1 0 1 X 3 6 = 3 6 3 6
```

There are more seats in
9 9 rows of 3 8 chairs.

1

99 rows of 38 chairs
101 rows of 36 chairs

2

99 rows of 33 chairs
101 rows of 27 chairs

3

99 rows of 54 chairs
101 rows of 52 chairs

4

99 rows of 23 chairs
101 rows of 17 chairs

5

99 rows of 82 chairs
101 rows of 78 chairs

6

99 rows of 31 chairs
101 rows of 28 chairs

7

99 rows of 67 chairs
101 rows of 54 chairs

 Write the difference between the total numbers of chairs for each pair.

Explore

Make five different 2-digit numbers, e.g. 76, 34, 85, 51, 12.

Multiply each number by 101. What do you notice?

Multiply each number by 1001. What do you notice?

Repeat for some 3-digit numbers.

Nine children wrote a multiplication with an answer close to 4000.

Write which three were the closest.

1  Gary
48 × 82

2 Greg
102 × 39

3 Tanya
47 × 83

4 Peta
52 × 78

5 Jamal
98 × 42

6 Katy
97 × 44

7 Tim
51 × 79

8 Sara
103 × 37

9 Amar
99 × 41

e Write the difference between each answer and 4000.

Problems

10 Wallpaper is **48 cm** wide.

Amelia uses **18** strips to wallpaper one side of her hall.

How many metres long is Amelia's hall?

11 A shop sells **24** stereos. Each stereo costs **£98·99**.

£98·99

How much money does the shop make?

12 Dotty has **101** dalmatians.

Every day each dalmatian eats **250 g** of dog food.

SPOT

How many kilograms of dog food does Dotty need to feed her dogs for one week?

Copy and complete each line.

Use doubling to help you.

1. 3 × 8 = 24
    3 × 16 = 48
    3 × 32 = 96

1   3 × 8 =   →   3 × 16 =   →   3 × 32 =

2   5 × 7 =   →   5 × 14 =   →   5 × 28 =

3   4 × 9 =   →   4 × 18 =   →   4 × 36 =

4   7 × 6 =   →   7 × 12 =   →   7 × 24 =

5   3 × 7 =   →   3 × 14 =   →   3 × 28 =

Write the cost of tickets for each group of children.

6. £8 × 13 = £104
    £16 × 13 = £208

Disco £16

Amusement Park £18

Football Match £28

6	13 children	7	24 children
8	11 children	9	31 children
10	12 children	11	20 children
12	35 children	13	23 children
14	17 children	15	21 children
16	33 children	17	42 children

Multiplying by doubling and halving

Copy and complete.

1  18×15

2 8×32

3 16×29

4 12×23

5 11×14

6 36×25

7 32×13

8 21×48

9 12×44

10 30×32

11 28×36

12 24×15

13 44×25

14 51×16

15 18×32

16 24×22

| pencils 24p each | rulers 35p each | rubbers 25p each | felt-tips 28p each | paint brushes 32p each |

Write the cost of these items.

Use doubling and halving.

17.
$$15 \times 24 = 30 \times 12$$
$$= 60 \times 6$$
$$= 360$$
$$15 \times 24p = £3·60$$

17

15 pencils

18

26 rulers

19

45 felt-tips

20

25 paint brushes

21

28 rubbers

22

42 rulers

23

75 pencils

24

34 rubbers

25

35 felt-tips

A factory takes 25 seconds to pack a box of chocolates.

Write how many seconds it takes to pack each set.

```
1. 24 × 100 = 2400
   24 × 50 = 1200
   24 × 25 = 600 seconds
```

1  24 boxes

2 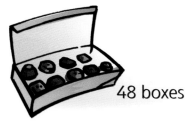 36 boxes

3  18 boxes

4 12 boxes

5  48 boxes

6 28 boxes

7 96 boxes

8 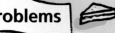 34 boxes

9  26 boxes

Write each time in minutes and seconds.

```
1a. 600 seconds = 10 minutes
```

10 A train has **25** carriages. Each carriage has **42** seats. If **1134** people get on the train, how many have to stand?

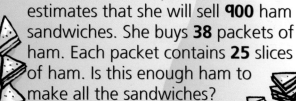

11 The sandwich shop owner estimates that she will sell **900** ham sandwiches. She buys **38** packets of ham. Each packet contains **25** slices of ham. Is this enough ham to make all the sandwiches?

Problems

12 Linacre School needs some new books for Year 6. The books cost **£12·20** each. There are **24** children in the class and the teacher needs a book too. What is the total cost?

13 Fastfoot United Football Club has **25** players. The kit for each player costs **£39**. How much change is there from **£1000**?

Finding fractions of quantities

Write the number of stickers each child buys.

1

$\frac{1}{6}$

2 $\frac{1}{5}$

3 $\frac{1}{9}$

4 $\frac{1}{10}$

5 $\frac{1}{4}$

6 $\frac{1}{8}$

7 $\frac{1}{7}$

Copy and complete.

8 $\frac{7}{8}$ of 32

8. $\frac{1}{8}$ of 3 2 = 4

$\frac{7}{8}$ of 3 2 = 2 8

9 $\frac{3}{5}$ of 25

10 $\frac{4}{7}$ of 28

11 $\frac{2}{3}$ of 60

12 $\frac{3}{10}$ of 140

13 $\frac{3}{4}$ of 44

14 $\frac{5}{6}$ of 180

15 $\frac{8}{9}$ of 72

16 $\frac{4}{11}$ of 121

Finding fractions of quantities

Each stall at the village fête gives $\frac{4}{10}$ of the money it makes to charity.

Write how much goes to charity for each.

1. $\frac{1}{10}$ of £125 = £12·50

   $\frac{4}{10}$ of £125 = £50

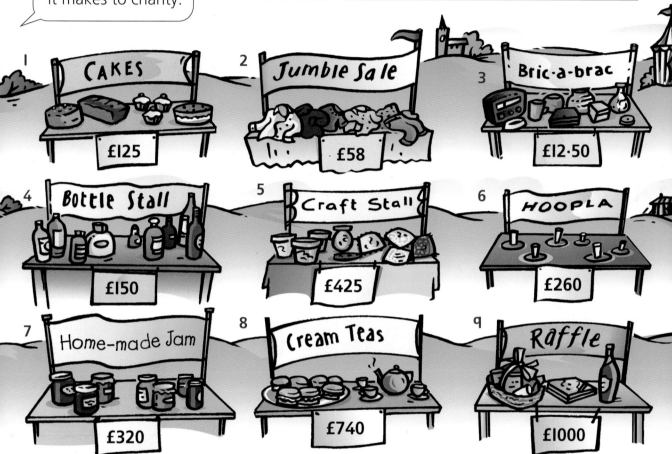

1   CAKES   £125

2   Jumble Sale   £58

3   Bric-a-brac   £12·50

4   Bottle Stall   £150

5   Craft Stall   £425

6   HOOPLA   £260

7   Home-made Jam   £320

8   Cream Teas   £740

9   Raffle   £1000

e How much money goes to charity altogether?

Copy and complete.

10. $\frac{1}{5}$ of 40 = 8

   $\frac{4}{5}$ of 40 = 32

10   $\frac{4}{5}$ of 40

11   $\frac{7}{8}$ of 48

12   $\frac{2}{9}$ of 81

13   $\frac{2}{7}$ of 77

14   $\frac{3}{4}$ of 28

15   $\frac{2}{3}$ of 90

16   $\frac{5}{6}$ of 120

17   $\frac{3}{8}$ of 240

18   $\frac{7}{10}$ of 500

Each child owes a fraction of their pocket money to Mum.

Write how much each child owes.

I. $\frac{1}{8}$ of £3·20 = 40p

$\frac{3}{8}$ of £3·20 = £1·20

1  £3·20 owes $\frac{3}{8}$

2 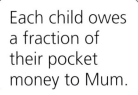 £4·00 owes $\frac{4}{5}$

3  £4·80 owes $\frac{3}{4}$

4  £1·70 owes $\frac{4}{10}$

5  £2·80 owes $\frac{2}{7}$

6  £5·60 owes $\frac{3}{8}$

7  £2·40 owes $\frac{5}{12}$

8  £5 owes $\frac{2}{25}$

9  £4·50 owes $\frac{7}{9}$

10 Ahmed has collected **56** monster cards.

He swaps **8** with his friend for some sweets.

He gives **8** more to his sister.

What fraction of the cards does he keep?

Problems

11 Gilly has **24 m** of ribbon. She gives $\frac{3}{8}$ to her mum and $\frac{1}{6}$ to her sister.

How many metres does she have left?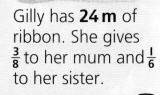

She gives $\frac{1}{2}$ of the remaining ribbon to her friend.

How much ribbon does she have now?

12 There are **120** dog biscuits in a packet. Chloe shares the biscuits between her **3** dogs.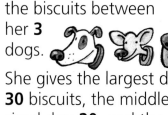

She gives the largest dog **30** biscuits, the middle-sized dog **20**, and the smallest dog **15**.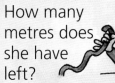

What fraction of the packet does each dog get?

How many biscuits are left in the packet?

What fraction is this?

Write five fractions which are equivalent to each of these.

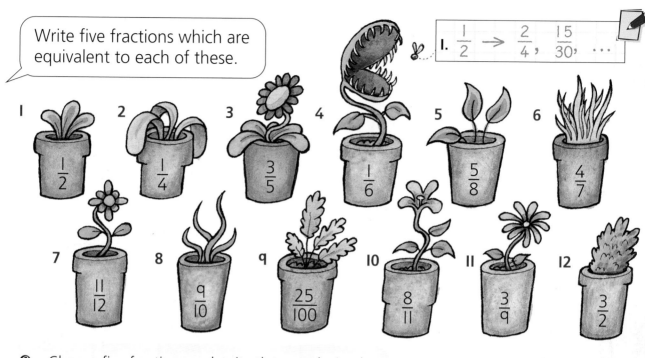

I. $\dfrac{1}{2} \rightarrow \dfrac{2}{4}, \dfrac{15}{30}, \cdots$

1 $\dfrac{1}{2}$

2 $\dfrac{1}{4}$

3 $\dfrac{3}{5}$

4 $\dfrac{1}{6}$

5 $\dfrac{5}{8}$

6 $\dfrac{4}{7}$

7 $\dfrac{11}{12}$

8 $\dfrac{9}{10}$

9 $\dfrac{25}{100}$

10 $\dfrac{8}{11}$

11 $\dfrac{3}{9}$

12 $\dfrac{3}{2}$

ℯ Choose five fractions and write them as decimals.

Write the simplest equivalent fraction for each.

13. $\dfrac{5}{20} = \dfrac{1}{4}$

1	2	3	4	5	6	7	8	9	10
2	4	6	8	10	12	14	16	18	20
3	6	9	12	15	18	21	24	27	30
4	8	12	16	20	24	28	32	36	40
5	10	15	20	25	30	35	40	45	50
6	12	18	24	30	36	42	48	54	60
7	14	21	28	35	42	49	56	63	70
8	16	24	32	40	48	56	64	72	80
9	18	27	36	45	54	63	72	81	90
10	20	30	40	50	60	70	80	90	100

13 $\dfrac{5}{20}$

14 $\dfrac{3}{15}$

15 $\dfrac{6}{18}$

16 $\dfrac{10}{80}$

17 $\dfrac{7}{63}$

18 $\dfrac{9}{72}$

19 $\dfrac{24}{64}$

20 $\dfrac{27}{90}$

21 $\dfrac{16}{36}$

22 $\dfrac{30}{45}$

ℯ Write two other fractions which are equivalent to each.

Equivalent fractions

Write the simplest equivalent fraction for each.

1 $\dfrac{27}{36}$

$$1. \quad \dfrac{27}{36} = \dfrac{3}{4}$$

2 $\dfrac{15}{20}$

3 $\dfrac{4}{14}$

4 $\dfrac{8}{20}$

5 $\dfrac{6}{9}$

6 $\dfrac{12}{42}$

7 $\dfrac{27}{45}$

8 $\dfrac{70}{100}$

9 $\dfrac{16}{36}$

Write fractions in their simplest forms.

Write the fraction of superheroes that:

$$10. \quad \dfrac{16}{24} = \dfrac{2}{3}$$

10 have a cloak

11 have a mask

12 have a belt

13 do not have a mask

14 do not have a belt or a mask

15 have a mask and a cloak

16 have a belt, but no cloak

17 have a cloak, but no mask

Equivalent fractions

Each child spends a fraction of their savings.

Write how much each child spends.

1. $\frac{6}{9} = \frac{2}{3}$

$\frac{1}{3}$ of £30 = £10

$\frac{2}{3}$ of £30 = £20

1
$\frac{6}{9}$ of £30

2
$\frac{6}{15}$ of £40

3
$\frac{15}{20}$ of £24

4
$\frac{8}{16}$ of £4

5
$\frac{24}{32}$ of £16

6
$\frac{32}{40}$ of £10

7
$\frac{9}{21}$ of £7

8
$\frac{36}{81}$ of £18

Explore

Use number cards 0 to 9.

Use four cards to make this correct.

$\frac{\square}{\square}$ of $\square\square = 8$

How many different ways of arranging the cards can you find?

$\frac{1}{3}$ of 24 = 8

Converting fractions

Convert each fraction into twenty-fourths.

Copy the number line and position the fractions on it.

a. $\frac{1}{2} = \frac{12}{24}$

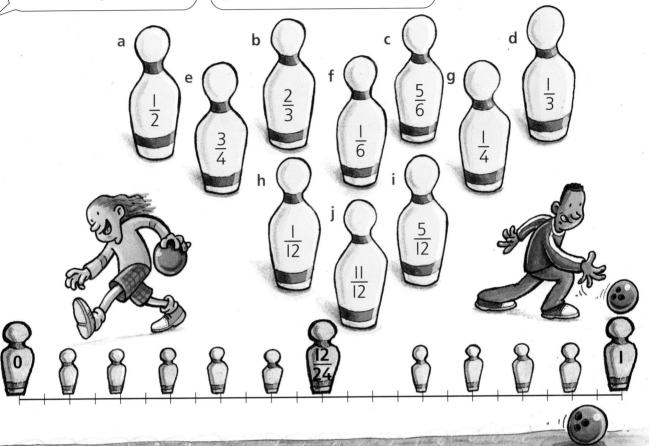

a $\frac{1}{2}$

e $\frac{3}{4}$

b $\frac{2}{3}$

f $\frac{1}{6}$

c $\frac{5}{6}$

g $\frac{1}{4}$

d $\frac{1}{3}$

h $\frac{1}{12}$

j $\frac{11}{12}$

i $\frac{5}{12}$

0    $\frac{12}{24}$ 1

Compare each pair of fractions by converting them to fractions with a common denominator.

Write '<' or '>' between each pair.

l. $\frac{3}{5} = \frac{9}{15}$

$\frac{2}{3} = \frac{10}{15}$

$\frac{3}{5} < \frac{2}{3}$

1 $\frac{3}{5}$ $\frac{2}{3}$

2 $\frac{2}{3}$ $\frac{3}{4}$

3 $\frac{3}{4}$ $\frac{4}{5}$

4 $\frac{5}{6}$ $\frac{7}{9}$

5 $\frac{3}{4}$ $\frac{5}{7}$

6 $\frac{5}{12}$ $\frac{3}{7}$

7 $\frac{4}{5}$ $\frac{5}{9}$

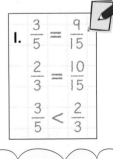

Find equivalent fractions with the same denominator.

29

Ordering fractions

Write each set of fractions so that they have the same denominator.

Write the fractions in order.

I. $\frac{2}{3} = \frac{40}{60}$

$\frac{3}{5} = \frac{36}{60}$

$\frac{3}{4} = \frac{45}{60}$

$\frac{3}{5} < \frac{2}{3} < \frac{3}{4}$

1
$\frac{2}{3}$ $\frac{3}{5}$ $\frac{3}{4}$

2
$\frac{2}{7}$ $\frac{1}{3}$ $\frac{8}{21}$

3
$\frac{5}{9}$ $\frac{2}{3}$ $\frac{11}{18}$

4
$\frac{1}{4}$ $\frac{5}{12}$ $\frac{2}{3}$

5
$\frac{3}{5}$ $\frac{1}{2}$ $\frac{4}{7}$

6
$\frac{3}{4}$ $\frac{4}{5}$ $\frac{7}{10}$

7
$\frac{1}{2}$ $\frac{7}{12}$ $\frac{5}{6}$

8
$\frac{3}{7}$ $\frac{2}{3}$ $\frac{4}{9}$

9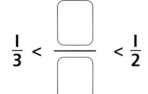
$\frac{1}{2}$ $\frac{1}{3}$ $\frac{5}{11}$

10
$\frac{9}{10}$ $\frac{3}{4}$ $\frac{11}{12}$

11
$\frac{2}{3}$ $\frac{4}{5}$ $\frac{5}{8}$

℮ Write a smaller fraction for each set.

Explore

How many fractions between $\frac{1}{3}$ and $\frac{1}{2}$ with denominators less than 20 can you find?

$\frac{1}{3} < \frac{\boxed{}}{\boxed{}} < \frac{1}{2}$

$\frac{1}{3} < \frac{9}{19} < \frac{1}{2}$

Finding the mid-point between fractions

Write the mid-point between each pair of fractions.

I. $2\frac{3}{7} = 2\frac{6}{14}$

$2\frac{4}{7} = 2\frac{8}{14}$

mid-point: $2\frac{7}{14} = 2\frac{1}{2}$

1 $2\frac{3}{7}$ $2\frac{4}{7}$

2 $3\frac{3}{5}$ $3\frac{4}{5}$

3 $1\frac{1}{3}$ $1\frac{2}{3}$

4 $2\frac{4}{7}$ $2\frac{5}{7}$

5 $1\frac{3}{8}$ $1\frac{5}{8}$

6 $4\frac{7}{9}$ $4\frac{8}{9}$

7 $3\frac{6}{11}$ $3\frac{7}{11}$

8 $5\frac{13}{15}$ $5\frac{14}{15}$

9 $6\frac{4}{5}$ $7\frac{1}{5}$

Problems

10 $\frac{2}{5}$ of a number is $\frac{3}{4}$ of **16**. What is the number?

11 $\frac{3}{8}$ of a number is $\frac{1}{3}$ of **18**. What is the number?

12 $\frac{4}{7}$ of a number is $\frac{1}{2}$ of **24**. What is the number?

13 $\frac{8}{15}$ of a number is $\frac{1}{8}$ of **64**. What is the number?

14 $\frac{9}{10}$ of a number is $\frac{6}{7}$ of **63**. What is the number?

15 $\frac{11}{12}$ of a number is $\frac{2}{3}$ of **330**. What is the number?

Adding to make 10 and 100

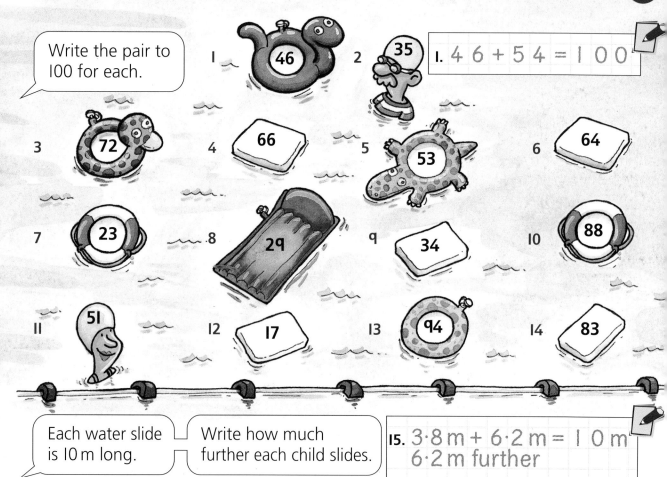

Write the pair to 100 for each.

1. $46 + 54 = 100$

1 _46_
2 _35_
3 _72_
4 _66_
5 _53_
6 _64_
7 _23_
8 _29_
9 _34_
10 _88_
11 _51_
12 _17_
13 _94_
14 _83_

Each water slide is 10 m long.

Write how much further each child slides.

15. $3·8 m + 6·2 m = 10 m$
    6·2 m further

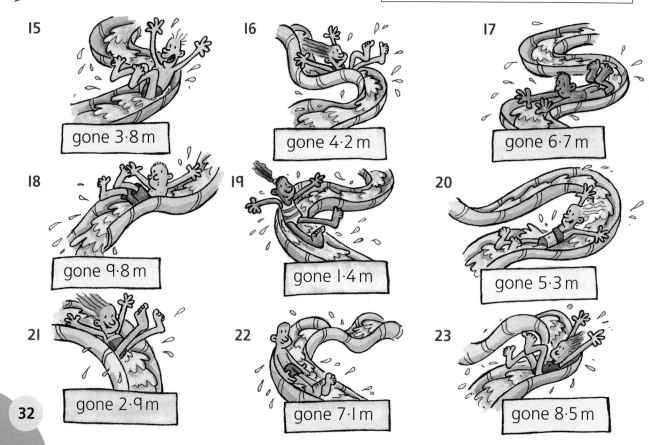

15 gone 3·8 m

16 gone 4·2 m

17 gone 6·7 m

18 gone 9·8 m

19 gone 1·4 m

20 gone 5·3 m

21 gone 2·9 m

22 gone 7·1 m

23 gone 8·5 m

Find how many people get off the bus at the last stop.

1

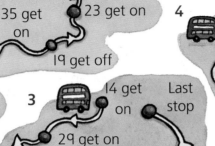

28 get on

Last stop

35 get on

23 get on

19 get off

I. $2\,8 + 3\,5 + 2\,3 - 1\,9 = 6\,7$

4

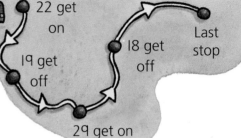

22 get on

Last stop

19 get off

18 get off

29 get on

2

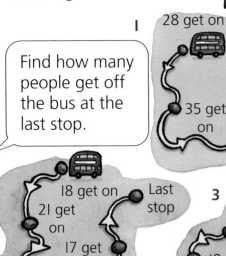

18 get on

Last stop

21 get on

17 get off

...get ...ff

19 get on

3

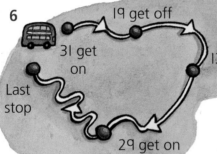

14 get on

Last stop

29 get on

18 get off

19 get on

7 get off

5

24 get on

Last stop

9 get off

17 get on

12 get off

6

19 get off

31 get on

13 get on

Last stop

29 get on

7

29 get off

38 get on

20 get off

11 get on

Last stop

---

8 A film is **87** minutes long. The director decides to cut **29** minutes of the film. The titles and credits add another **7** minutes.

How many more minutes of film must the director cut to fit it on a **55**-minute video tape?

Problems

9 Jamal and Harriet grow **48** tomato plants.

19 are eaten by rabbits. They sell **27** plants.

They share the remaining plants between them. How many tomato plants do they each get?

10 Ricky travels by train from Oxford to Birmingham. The journey usually takes **89** minutes.

A broken bridge delays the train by **27** minutes, but the train driver goes faster to make up **11** minutes.

If Ricky leaves at **10 a.m.**, what time does he arrive?

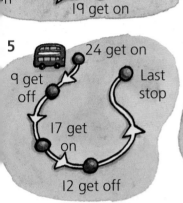

Find how many pages each child has read. ?

1

read 87 pages
read 29 pages
read 37 pages

2
read 111 pages
turned back 19 pages
read 37 pages
read 25 pages

3

read 204 pages
turned back 87 pages
read 29 pages
turned back 35 pages
read 42 pages

Write how many grams each baby bear weighs now. ?

4

weighed 347 g
lost 98 g
gained 117 g
gained 219 g

5

weighed 399 g
lost 86 g
gained 134 g
gained 122 g

6

weighed 407 g
lost 76 g
gained 121 g
lost 46 g
gained 237 g

Write how much money each child has now. ?

7
BANK
saved £2·24
spent 99p
earned £4·50
spent £1·65

8
BANK
saved £3·01
spent £0·78
spent £1·57
earned £3·50

9

BANK
saved £5·84
earned £2·50
spent £1·99
earned £2·76

Subtracting

Subtract by counting on.

I.

$$377 \qquad 434$$
$$23 \searrow 400 \nearrow 34$$
$$434 - 377 = 57$$

1   434 – 377   **2**   521 – 466

3   118 – 77   **4**   204 – 188

5   312 – 254   **6**   723 – 686

7   637 – 552   **8**   1049 – 973

Subtract by taking away.

q.

$$202 - 2 = 200$$
$$200 - 14 = 186$$
$$202 - 16 = 186$$

q   202 – 16   **10**   3014 – 25

11   5024 – 25   **12**   6006 – 18

13   412 – 18   **14**   121 – 32

15   2013 – 24   **16**   907 – 38

Subtract by rounding, then adjusting.

I7.

$$486 - 240 = 246$$
$$486 - 239 = 247$$

17   486 – 239   **18**   557 – 129

19   666 – 339   **20**   841 – 219

21   755 – 549   **22**   381 – 202

23   472 – 101   **24**   657 – 298

Choose how to subtract.

Write which method you use.

25   646 – 554   **26**   803 – 17

27   574 – 219   **28**   631 – 589

29   7008 – 26   **30**   367 – 229

31   304 – 268   **32**   119 – 78

Write how far each lorry must go to reach half-way.

1
216 km
gone 98 km

2
550 km
gone 177 km

3
448 km
gone 188 km

4
638 km
gone 279 km

5
326 km
gone 102 km

6
812 km
gone 386 km

7
144 km
gone 25 km

Find how many children have school dinners.

8
214 children
178 eat sandwiches
0 go home

q
331 children
184 eat sandwiches
48 go home

10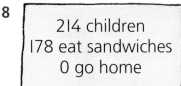
277 children
102 eat sandwiches
73 go home

11
405 children
278 eat sandwiches
27 go home

12
245 children
166 eat sandwiches
41 go home

13
426 children
289 eat sandwiches
38 go home

14
371 children
203 eat sandwiches
52 go home

Subtracting

Each car park has a different number of spaces.

Write how many empty spaces there are at 11 o'clock.

I.
$$5 \times 16 = 80$$
$$80 - 29 - 34 + 27 = 44$$
44 empty spaces

5 rows of 16 spaces

1
9 a.m.	29 cars in
10 a.m.	34 cars in
11 a.m.	27 cars out

2
9 a.m.	13 cars in
10 a.m.	38 cars in
11 a.m.	24 cars out

3
9 a.m.	31 cars in
10 a.m.	19 cars out
11 a.m.	23 cars in

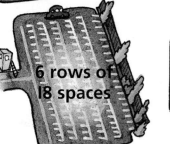

6 rows of 18 spaces

4
9 a.m.	18 cars in
10 a.m.	25 cars in
11 a.m.	33 cars out

5
9 a.m.	32 cars in
10 a.m.	17 cars in
11 a.m.	46 cars out

6
9 a.m.	41 cars in
10 a.m.	37 cars in
11 a.m.	30 cars in

9 rows of 22 spaces

7
9 a.m.	19 cars in
10 a.m.	51 cars in
11 a.m.	76 cars in

8
9 a.m.	28 cars in
10 a.m.	14 cars out
11 a.m.	39 cars in

9
9 a.m.	67 cars in
10 a.m.	58 cars in
11 a.m.	125 cars out

Explore

$$182 - 49 = 133$$

$$209 - 13 = 196$$

$$328 - 277 = 51$$

Write a word problem for each of these calculations.

Use your own numbers to write a calculation and word problem. Choose your subtraction method.

Swap your problems with a friend and solve them.

182 − 49 = 133

A shop has 182 CDs. It sells....

Multiples

Write which of these numbers are:

1	2	3	4	5	6	7	8	9	10
11	12	13	14	15	16	17	18	19	20
21	22	23	24	25	26	27	28	29	30
31	32	33	34	35	36	37	38	39	40
41	42	43	44	45	46	47	48	49	50

1 multiples of 7

2 multiples of 9

3 multiples of 13

4 common multiples of 3 and 4

1. 7, 14, 21, 28, …

5 multiples of both 2 and 3

6 common multiples of 5 and 6

7 multiples of both 4 and 5

8 not multiples of 2, 3 or 4

9 common multiples of 3 and 5

10 common multiples of 4 and 8

ℓ Write the percentage of the numbers that are these multiples.

24 9 18 36 50 11 48 16 25 13 20 21

Write which of these numbers are:

11. 24, 18, 36, 50, 48, 16, 20

11 multiples of 2

12 multiples of 3

13 multiples of 4

14 common multiples of 2 and 3

15 common multiples of 3 and 4

16 common multiples of 3 and 7

17 common multiples of 2 and 5

Common multiples

Write a number that is a multiple of:

1 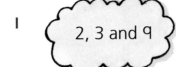 2, 3 and 9

I. 5 4

2 2, 3 and 13

3 2, 9 and 18

4 2, 3, 4 and 5

5 6, 8 and 15

6 12, 15 and 20

7 4, 6 and 9

8 3, 4 and 18

9 2, 3 and 21

10 4, 5, 6 and 10

Write the lowest common multiple of each set.

Ia. 1 8

Find the lowest common denominator for each pair of fractions.

Use this to compare the fractions.

II. 2 1

$$\frac{9}{21} < \frac{14}{21}$$

$$\frac{3}{7} < \frac{2}{3}$$

11 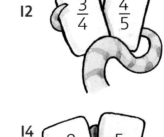 $\frac{3}{7}$ $\frac{2}{3}$

12 $\frac{3}{4}$ $\frac{4}{5}$

13  $\frac{5}{6}$ $\frac{4}{7}$

14 $\frac{8}{9}$ $\frac{5}{6}$

15 $\frac{5}{12}$ $\frac{2}{5}$

16  $\frac{7}{10}$ $\frac{11}{25}$

17 $\frac{9}{17}$ $\frac{3}{4}$

18  $\frac{7}{12}$ $\frac{17}{30}$

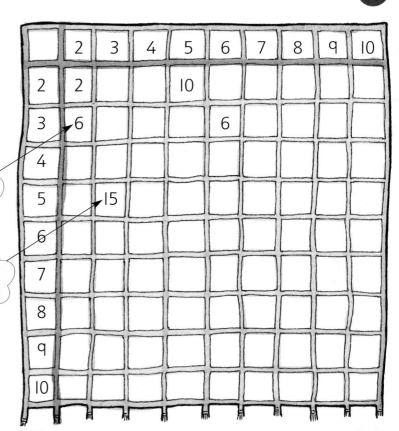

Copy and complete the table to show the lowest common multiple of each row and column number.

lowest common multiple of 2 and 3

lowest common multiple of 3 and 5

	2	3	4	5	6	7	8	9	10
2	2			10					
3	6				6				
4									
5		15							
6									
7									
8									
9									
10									

Problems

1   I am a multiple of 12.
I am a multiple of 9.
I am less than 50.
Who am I?

2   I am a common multiple of 2, 4 and 7.
I have a digit total of 11.
Who am I?

3   I am a multiple of 8.
I am a common multiple of 2 and 3.
One of my digits is double the other.
I am less than 30.
Who am I?

 Explore

Use the number cards shown.

Explore sets of three cards which have lowest common multiples of 30 or more.

How many can you find?

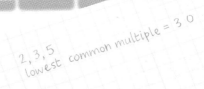

2 3 4 5 6 7 8

2, 3, 5
lowest common multiple = 30

Divisibility

 Can these flowers be shared equally into 4 flower beds?

Write 'yes' or 'no'.

I. no

1   34 flowers

2   48 flowers

3   66 flowers

4   84 flowers

5   132 flowers

6   116 flowers

7   224 flowers

8   314 flowers

9   152 flowers

10   168 flowers

11   322 flowers

12   98 flowers

e   Window boxes hold 25 flowers. How many window boxes can be filled from each set of flowers? How many flowers are left over?

Write which numbers divide by 2, 4 or 8.

13. 8 6 divides by 2

13  86

14  420

15  214

16  152

17  78

18 108

19  92

20  146

21 262

22 188

23 324

24 504

45

Write which numbers divide by 3.

1   462

1. 4 + 6 + 2 = 1 2
   1 2 divides by 3
   4 6 2 divides by 3

2   558

3   662

4   435

5   1008

6   990

7   738

8   444

9   6021

10   7398

11   9045

Write which numbers divide by 6.

1a. 4 6 2 divides by 3
    4 6 2 is even
    4 6 2 divides by 6

℮ Write which numbers divide by 9.

Can these people be put into 6-a-side teams exactly?

Write 'yes' or 'no'.

12. no

12   112 people

13   224 people

14   360 people

15   144 people

16   450 people

17   524 people

18   196 people

19   252 people

Can these people be put into 9-a-side teams exactly?

12a. no

℮ For those which can be put exactly into teams, write how many teams.

Divisibility

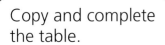

Copy and complete the table.

	÷ 2	÷ 3	÷ 4	÷ 5	÷ 6	÷ 8	÷ 9	÷ 10	÷ 25
60	✓								
720									
945									
7644									
1000									✓
2508		✓							
1475									

1

I am divisible by 4.
I am divisible by 3.
I am a 2-digit number.
Both my digits are
even. I am more than
60. Who am I?

Problems

2

I am divisible by 8.
I am not divisible by 3.
I am not divisible by 5.
I am between 35
and 85. I have one
odd and one even digit.
Who am I?

3

I am divisible by 9.
I have two digits.
My tens digit is double
my units digit.
Who am I?

Explore

Try this test for divisibility by 7 for 3-digit numbers.

* Double the first digit.
* Add this to the last two digits.
* Is the answer a multiple of 7?
* If yes, then the number is divisible by 7.

245 148 364 295 259

Which of these numbers are divisible by 7?

Can you find some 4-digit numbers that
are divisible by 7?

Can you find a test for divisibility by 14?

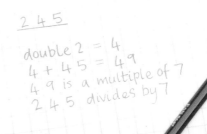

2 4 5

double 2 = 4
4 + 45 = 49
49 is a multiple of 7
2 4 5 divides by 7

Each fence has 10 panels.

Write the total length of each fence.

I. $10 \times 3.8\,m = 38\,m$

1

3·8 m

2
4·4 m

3
1·5 m

4

3·5 m

5
1·2 m

6

2·8 m

7
3·1 m

8
0·9 m

q
0·7 m

10
4·1 m

11

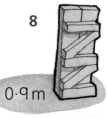
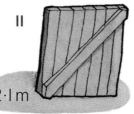
2·1 m

12

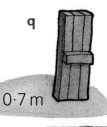

3·7 m

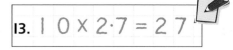

Write the missing numbers.

13. $10 \times 2.7 = 27$

13 $10 \times 2.7 = $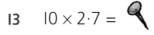

14  $\times 1.3 = 13$

15 $10 \times $ = 95

16 $0.8 \times $ = 8

17 $10 \times 3.4 = $

18 $10 \times $ = 134

19 $10 \times 21.3 = $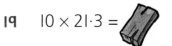

20 $\times 10 = 60$

21 $\times 19.7 = 197$

22 $10 \times 43.8 = $

23 $10 \times $ = 40

24 $10 \times 12.5 = $

25 $\times 61.4 = 614$

48

Multiplying decimals

 Write the missing numbers.

1   $10 \times 1.74 =$

I. $10 \times 1.74 = 17.4$

2   $10 \times$ $= 23.1$

3   $100 \times 1.86 =$

4    $\times 1.92 = 19.2$

5    $\times 3.16 = 316$

6   $10 \times$ $= 27.3$

7   $100 \times 0.93 =$

8   $10 \times$  $= 7.6$

9   $100 \times 4.7 =$

10   $100 \times$  $= 209$

A shop orders 10 of each item. — Write the total cost.

II. $10 \times £3.26 = £32.60$

11    £3·26

12   £5·93

13   £9·34

14   £1·75

15   £6·84

16   £11·20

17   £4·82

18   £7·02

19   £21·30

20   £14·51

21   £7·39

22   £0·28

 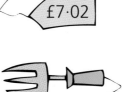 

Write the cost of 100 of each.

IIa. $100 \times £3.26 = £326$

Write the missing numbers.

1  × 2·3 = 230

I. 1 0 0 × 2·3 = 2 3 0

2 10 ×  = 15

3 10 ×  = 34

4 100 ×  = 760

5 100 ×  = 480

6  × 31·2 = 312

7  × 4·23 = 423

8 100 ×  = 675

9 10 ×  = 80·9

10 100 ×  = 938

11  × 16·45 = 1645

12 10 ×  = 93·8

13 100 ×  = 895

14  × 67 = 6700

15 1000 ×  = 730

16 1000 ×  = 210

Problems

17 Stephen cycles **3·7 km** to work and the same distance home **5** days a week.

How far does he cycle in a fortnight? How far in **20** weeks? How far in a year if he has **4** weeks holiday?

18 A beanstalk is **3 mm** high. Each day it grows to **10** times its original height that day. How many days will it take until it is **3 km** high?

Explore

Multiply 2·1 by 10, 100, 1000, 20 and 200.

Explore quick ways of multiplying other decimal numbers by 10, 100, 1000, 20 and 200.

How quickly can you find the answers mentally?

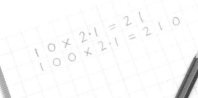

> These are the total distances each child travels to and from school each week.

> Write how far each child lives from school.

1. 4 7·3 km ÷ 1 0
      = 4·7 3 km

1

47·3 km

2

61·4 km

3

9·8 km

4

11 km

5

3·5 km

6

16·3 km

7

75 km

8

8 km

9

119·5 km

℮ How far does each child travel in 100 school days?

> Copy and complete.

10 $14 \div 10 =$

10. 1 4 ÷ 1 0 = 1·4

11    $\div 10 = 0.173$

12 $6 \div$ $= 0.6$

13   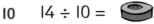 $\div 10 = 0.87$

14 $73 \div 10 =$

15 $7.0 \div 10 =$

16    $\div 10 = 0.975$

17 $0.3 \div 10 =$

18 $\div 10 = 1.76$

19 $1.1 \div 10 =$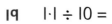

Dividing decimals

Write the height of each animal in metres.

1 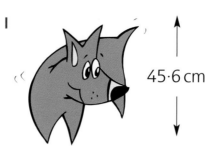 45·6 cm

I. 4 5·6 ÷ 1 0 0
= 0·4 5 6 m

2  28·5 cm

3  140·6 cm

4 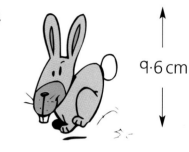 9·6 cm

5  230 cm

6  8·7 cm

7  51·6 cm

8 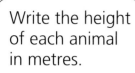 60·5 cm

9 98 cm

10 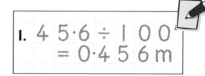 121 cm

Copy and complete.

11 36 ÷ 100 =

II. 3 6 ÷ 1 0 0 = 0·3 6

12 72·5 ÷ 100 =

13 ÷ 100 = 0·59

14 40·3 ÷ 100 =

15 146 ÷  = 1·46

16 4·6 ÷ 100 =

17  ÷ 100 = 0·07

18 0·8 ÷ 100 =

19  ÷ 100 = 0·077

20 96 ÷ 100 =

Dividing decimals

Write the missing numbers.

1 $3.8 \div$ 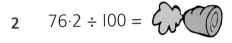 $= 0.38$

$$\text{1. } 3.8 \div 10 = 0.38$$

2 $76.2 \div 100 =$

3 $0.6 \div$ $= 0.06$

4 $0.04 \div$ $= 0.0004$

5 $\div 10 = 0.91$

6 $45.62 \div$ $= 4.562$

7 $\div 100 = 0.008$

8 $\div 100 = 0.075$

9 $96.3 \div 10 =$

10 $1.25 \div$ $= 0.0125$

11 $77.3 \div$ $= 0.773$

Problems

12 A bag of **100** balls weighs **8·3 kg**.

What is the weight of each ball in kilograms?

What is the weight of each ball in grams?

13 A group of **10** friends win a competition and share the prize of **£75·30**.

How much does each winner get?

How much more do they need to each have **£10**?

14 A sprinter runs a **100 m** race in **9·92** seconds.

What is his average time to run **10** metres?

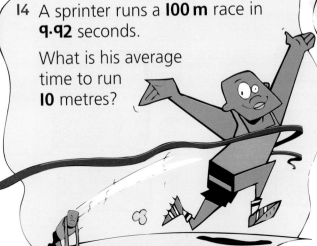

15 A barbecue stand **0·72 m** long is made from **10** bricks placed in a row.

What is the length of each brick in metres?

How many bricks are needed to make a barbecue stand which is just over **1 m** long?

53

Multiplying

The taxi companies charge different rates.

Write the cost of each journey.

1. $(7 \times 40)+(7 \times 5)= 3\ 1\ 5$
   280   35

7 miles cost £ 3·1 5

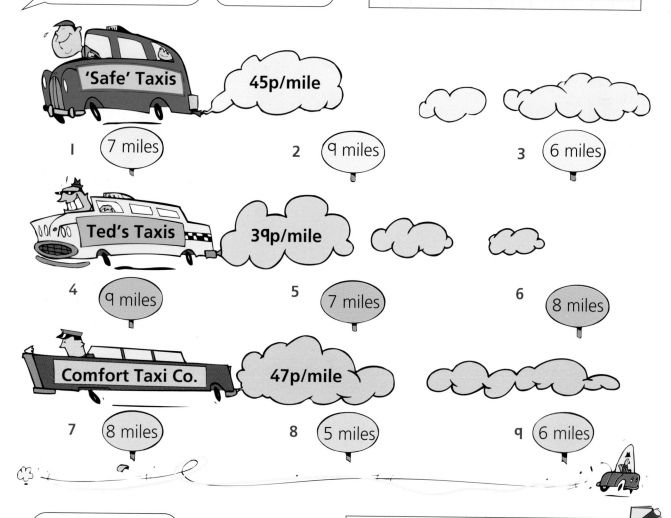

'Safe' Taxis — 45p/mile

1 7 miles   2 9 miles   3 6 miles

Ted's Taxis — 39p/mile

4 9 miles   5 7 miles   6 8 miles

Comfort Taxi Co. — 47p/mile

7 8 miles   8 5 miles   9 6 miles

Complete these multiplications.

10 4×73

10. $(4 \times 70)+(4 \times 3)= 2\ 9\ 2$
    280   12

11 3×42 12 5×35 13 7×61

14 8×33 15 6×18 16 9×75

17 4×83 18 7×29 19 8×46

Multiplying

Write the cost of each holiday.

1  Farmhouse **£317/week**

2 weeks

```
I. 2 × £300 = £600
   2 ×  £10 =  £20
   2 ×   £7 =  £14
   2 × £317 = £634
```

2  Cottage **£463/week**

2 weeks

3  Hotel **£542/week**

3 weeks

4  Caravan **£160/week**

4 weeks

5  Apartment **£285/week**

6 weeks

6  Camping **£125/week**

8 weeks

7  Canal boat **£665/week**

2 weeks

8  Guesthouse **£217/week**

3 weeks

9 Youth Hostel **£85/week**

5 weeks

10 Villa **£730/week**

3 weeks

℮ A travel agent gives a 10% discount on holidays of 3 or more weeks. Write the final cost of each holiday.

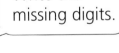

Write the missing digits.

11 $3 \times \bigcirc 6 = 13 \triangle$

12 $4 \times \bigcirc 2 = 8 \triangle$

13 $30 \triangle = 5 \times \bigcirc 1$

14 $4 \times \bigcirc 3 = 21 \triangle$

15 $16 \triangle = 6 \times \bigcirc 8$

Explore

Write a 2-digit multiple of 11. Multiply it by 9.

Repeat for 6 different numbers. What do you notice? Write about the pattern.

Repeat for multiplying 2-digit multiples of 11 by 8. Write about the pattern.

Explore for multiplying by other 1-digit numbers.

66 × 9 = 594
44 × 9 =

Eight children have written a multiplication.

Estimate the answers and put them in order, largest to smallest.

A
Anna
3 × 197

B
Beryl
8 × 76

C
Ceri
4 × 158

D
Davinder
9 × 65

E
Eilidh
5 × 126

F
Fiona
7 × 83

G
Gary
6 × 95

H
Hattie
3 × 172

Complete each multiplication.

Write the correct order.

Compare the correct order with your estimated order.

F, A, ...

Problems

I The Abbott family have booked a holiday which costs **£84** per adult. Children go half-price.

What is the total cost for **3** adults and **4** children?

How much change will there be from **£500**?

2 Karen goes swimming every week.

She has not missed a swimming session for **7** years.

How many times has she been swimming?

3 Tanya is **4** years old today.

How many days old is she?

How many weeks until Tanya is **1510** days old?

4 The Barratt family have bought a television.

They can either pay **£147** per month for **6** months, or **£93** per month for **9** months.

Which is cheapest? What is the difference in total cost?

Multiplying by I-digit numbers

The shop sale finished today.

Write how much money the shop made for each item.

1. £4000

$$
\begin{array}{r}
£436 \\
\times \quad 8 \\
\hline
48 \\
240 \\
3200 \\
\hline
£3488 \\
\end{array}
$$

8 × 6
8 × 30
8 × 400

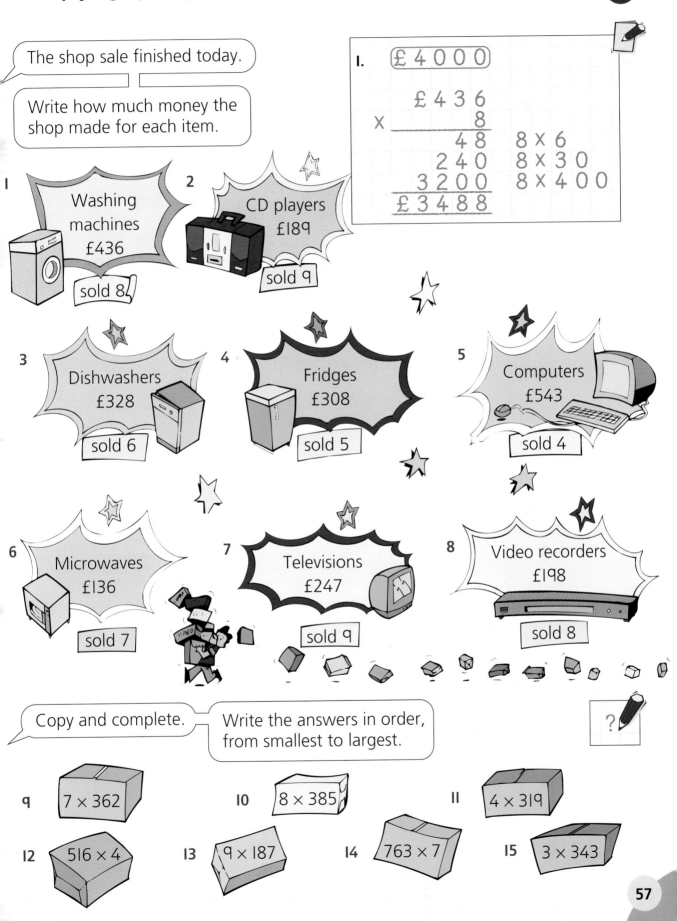

1 Washing machines £436 — sold 8

2 CD players £189 — sold 9

3 Dishwashers £328 — sold 6

4 Fridges £308 — sold 5

5 Computers £543 — sold 4

6 Microwaves £136 — sold 7

7 Televisions £247 — sold 9

8 Video recorders £198 — sold 8

Copy and complete. Write the answers in order, from smallest to largest.

9 7 × 362

10 8 × 385

11 4 × 319

12 516 × 4

13 9 × 187

14 763 × 7

15 3 × 343

In a maths quiz the contestants had to write a multiplication with an answer between 20 000 and 30 000.

Find which contestants are correct.

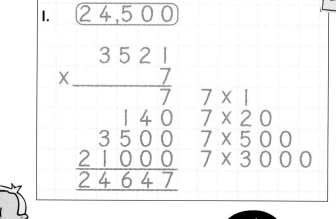

```
1.   (24,500)

        3 5 2 1
     x         7
             7      7 X 1
         1 4 0      7 X 2 0
       3 5 0 0      7 X 5 0 0
     2 1 0 0 0      7 X 3 0 0 0
     2 4 6 4 7
```

1 Danny
3521 × 7

2 Corin
4328 × 5

3 Hakki
9546 × 3

4 Ben
5314 × 4

5 Dionne
2789 × 5

6 Penny
7162 × 6

7 Trish
1958 × 8

8 Andrei
4629 × 4

q Mellina
6542 × 7

℮ Write the difference between each answer and 20 000.

Copy and complete.

10 4 × 3176

11 5 × 4826

```
10.     3 1 7 6
     x          4
     1 2 7 0 4
            3  2
```

12 4713 × q

13 6 × 4849

14 7 × 8320

15 5 × 4065

16 3906 × 4

Write the area of each worktop.

1 34 cm · 217 cm

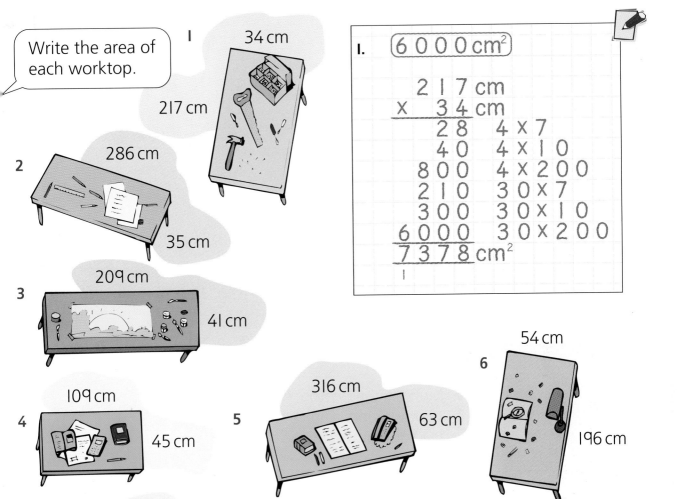

1. $\boxed{6000 \text{ cm}^2}$

```
      2 1 7  cm
  x    3 4  cm
      2 8      4 X 7
      4 0      4 X 1 0
    8 0 0      4 X 2 0 0
    2 1 0     3 0 X 7
    3 0 0     3 0 X 1 0
  6 0 0 0     3 0 X 2 0 0
  7 3 7 8  cm²
      1
```

2 286 cm · 35 cm

3 209 cm · 41 cm

4 109 cm · 45 cm

5 316 cm · 63 cm

6 54 cm · 196 cm

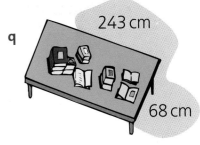

7 223 cm · 48 cm

8 146 cm · 34 cm

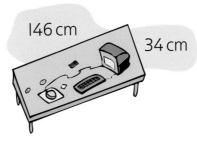

9 243 cm · 68 cm

Explore

2 3 4 5 6

Use the number cards shown.

Make a 2-digit number and a 3-digit number.

Multiply them together.

Explore the largest and smallest possible answers.

```
      2 3 4
    x   5 6
        2 4     6 X 4
      1 8 0     6 X 3 0
    1 2 0 0     6 X 2 0 0
      2 0 0     5 0 X 4
    1 5 0 0     5 0 X 3 0
  1 0 0 0 0     5 0 X 2 0 0
  1 3 1 0 4
        1 1
```

Complete each multiplication.

Write two division facts for each.

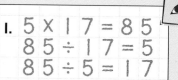
I. $5 \times 17 = 85$
$85 \div 17 = 5$
$85 \div 5 = 17$

1   $5 \times 17 =$ 2 $4 \times 3{\cdot}2 =$

3   $6 \times 75 =$ 4 $2{\cdot}8 \times 5 =$ 5 $20 \times 0{\cdot}3 =$

6   $25 \times 14 =$ 7 $32 \times 19 =$ 8 $4{\cdot}7 \times 30 =$

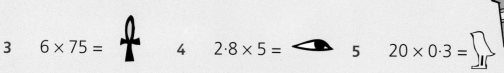

Copy and complete each triangle.

Multiply the bottom two numbers to give the top number.

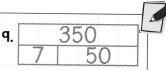

q.	350	
	7	50

q

7	50

10

4	30

11

40	6

12

7	20

13

240	
8	

14

350	
	5

15

100	
	4

16

7	25

17

225	
	25

Write four facts for each.

qa. $350 \div 7 = 50$
$350 \div 50 = 7$
$7 \times 50 = 350$
$50 \times 7 = 350$

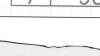

Multiplying and dividing

Clues!

$1.2 \times 3 = 3.6$

$0.9 \times 4 = 3.6$

$1.2 = 3 \times 0.4$

$2.7 = 0.9 \times 3$

$1.2 \times 4 = 4.8$

$3.6 = 2 \times 1.8$

Use the clues to solve these problems.

1 What is one third of 2·7?

2 What is three lots of 0·4?

3 What is double 1·8?

4 How many 0·9s make 2·7?

5 What is 2·7 divided by nine tenths?

6 What is one quarter of 3·6?

7 What is treble 1·2?

8 What is 4·8 divided by 4?

9 How many 0·4s make 12?

10 What is one half of 1·8?

11 What is 4 lots of 0·9?

12 What is three and six tenths divided by one and two tenths?

Copy and complete each circle.

Multiply the bottom two numbers to give the top number.

13

	2·4
4	0·6

13. circle: 4 | 0·6

14 circle: 1 / 5

15 circle: 4 / 8

16 circle: 2 / 0·4

e Write two divisions for each.

Multiplying and dividing

Use these facts to help solve the problems.

$5 \times 84 = 420$

$48 \div 3 \cdot 2 = 15$

$192 \div 16 = 12$

$2 \cdot 2 \times 15 = 33$

$6 \times 9 \cdot 5 = 57$

$57 \times 7 = 399$

Problems

1 Dad is saving to buy ferry tickets for a family of **5**. He has saved **£350**. The tickets cost **£84** each. How much more does he need?

2 Anne has **15** game boxes which are all the same size. They take up **33 cm** on her shelf. How much space do **6** game boxes take up?

3 Sammy is **16** years old. How many months until he has been alive for **200** months?

4 The weight of **6** identical marbles is **57 g**. What is the weight of **10** of the marbles?

5 A bucket holds **3·2 l** of water. How many buckets are needed to fill a tank which holds **96 l**?

6 A boat of **6** sailors sailed around the world in **57** weeks. How many days did the trip take?

 Explore

Use three of these numbers to make a multiplication or division fact.

Explore how many different multiplications and divisions there are using sets of three of these numbers.

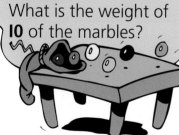

2   3   5   0·6   1·2   0·4   1·8

$2 \times 0 \cdot 6 = 1 \cdot 2$

Remainders

Complete each division.

Write the remainder as a fraction.

I. $43 \div 6 = 7\frac{1}{6}$

1 $43 \div 6 =$

2 $29 \div 4 =$

1	2	3	4	5	6	7	8	9	10
2	4	6	8	10	12	14	16	18	20
3	6	9	12	15	18	21	24	27	30
4	8	12	16	20	24	28	32	36	40
5	10	15	20	25	30	35	40	45	50
6	12	18	24	30	36	42	48	54	60
7	14	21	28	35	42	49	56	63	70
8	16	24	32	40	48	56	64	72	80
9	18	27	36	45	54	63	72	81	90
10	20	30	40	50	60	70	80	90	100

3 $31 \div 2 =$

4 $43 \div 5 =$

5 $73 \div 7 =$

6 $29 \div 3 =$

7 $64 \div 9 =$

8 $51 \div 6 =$

9 $33 \div 8 =$

10 $86 \div 9 =$

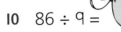

Write how many packs are needed.

11

41 crayons packs of 3

II. $41 \div 3 = 13\frac{2}{3}$

14 packs needed

12

75 crayons packs of 4

13

47 crayons packs of 3

14

81 crayons packs of 6

15

76 crayons packs of 5

16

86 crayons packs of 7

17

51 crayons packs of 4

18

75 crayons packs of 6

19

87 crayons packs of 9

20

93 crayons packs of 8

Petunias are sold in trays of 10. | Write how many trays are filled. | Write the remainder as a fraction and a decimal.

1 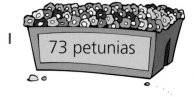 73 petunias

2 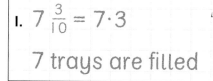 19 petunias

1. $7\frac{3}{10} = 7 \cdot 3$

7 trays are filled

3 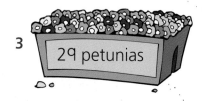 29 petunias

4  87 petunias

5 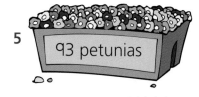 93 petunias

6 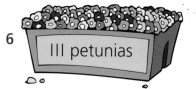 111 petunias

7  67 petunias

8 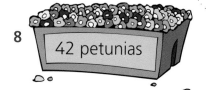 42 petunias

  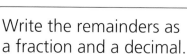

Tulip bulbs are sold in bags of 100. | Write how many bags are needed. | Write the remainders as a fraction and a decimal.

9  147 bulbs

10 359 bulbs

11 731 bulbs

9. $1\frac{47}{100} = 1 \cdot 47$

2 bags needed

12 867 bulbs

13 478 bulbs

14 209 bulbs

15 142 bulbs

16 790 bulbs

17 703 bulbs

18 84 bulbs

19 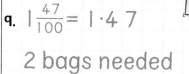 70 bulbs

20 1042 bulbs

21 1369 bulbs

22  2718 bulbs

23  9764 bulbs

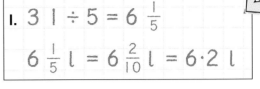

An equal amount of paint is poured into 5 buckets.

Write how much paint there is in each bucket.

Use fractions and decimals.

1  31 l

2  43 l

I. $31 \div 5 = 6\frac{1}{5}$

$6\frac{1}{5}$ l $= 6\frac{2}{10}$ l $= 6 \cdot 2$ l

3  76 l

4 89 l

5 38 l

6 47 l

7  29 l

8 36 l

Write which is larger in each pair.

Write each as a decimal.

?

9 $\frac{1}{10}$ of 47 $\frac{1}{5}$ of 23

10 $\frac{1}{4}$ of 35 $\frac{1}{100}$ of 860

11 $\frac{1}{2}$ of 13 $\frac{1}{5}$ of 32

12 $\frac{1}{4}$ of 70 $\frac{1}{10}$ of 176

13 $\frac{1}{100}$ of 94 $\frac{1}{2}$ of 19

14 $\frac{2}{5}$ of 50 $\frac{1}{4}$ of 90

15 $\frac{3}{10}$ of 61 $\frac{1}{2}$ of 76

16 $\frac{3}{5}$ of 42 $\frac{7}{10}$ of 55

Problems

17 **70** children are going to the swimming pool.
The minibus holds **8** children.
How many trips must it make?

18 Jake has **£20** to buy food for his party. Small pizzas cost **£3**, large pizzas cost **£4**. What different combinations can he buy for his party?

Mixed numbers and improper fractions

Write the number of slices.

1. $2\frac{3}{8} = \frac{19}{8}$

   1 9 slices

1   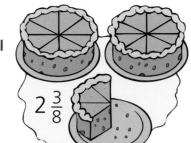 $2\frac{3}{8}$

2   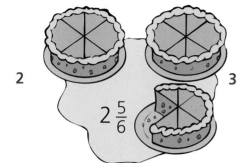 $2\frac{5}{6}$

3   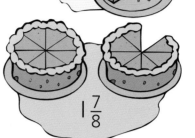 $1\frac{7}{8}$

4   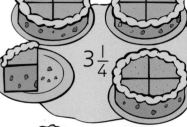 $3\frac{1}{4}$

5   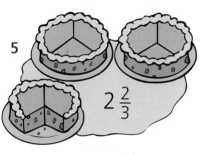 $2\frac{2}{3}$

6   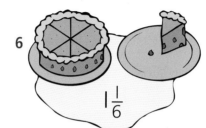 $1\frac{1}{6}$

7   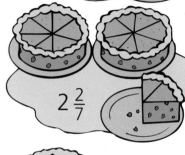 $2\frac{2}{7}$

8   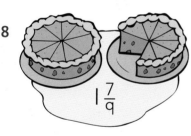 $1\frac{7}{9}$

9   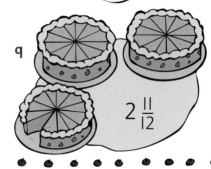 $2\frac{11}{12}$

10    $1\frac{7}{10}$

Write the number of fifths.

11   $3\frac{4}{5}$ 12 $1\frac{2}{5}$

II. $3\frac{4}{5} = \frac{19}{5}$

13   $2\frac{3}{5}$ 14 $3\frac{1}{5}$ 15 $2\frac{4}{5}$ 16 $3\frac{3}{5}$

17   $6\frac{1}{5}$ 18 $1\frac{4}{5}$ 19 $2\frac{2}{5}$ 20 $4\frac{2}{5}$

Mixed numbers and improper fractions

> The pattern on each tray is made with a different number of tiles.

> Write how many trays can be made.

I. $17 \div 8 = 2\frac{1}{8}$

17 tiles makes $2\frac{1}{8}$ trays.

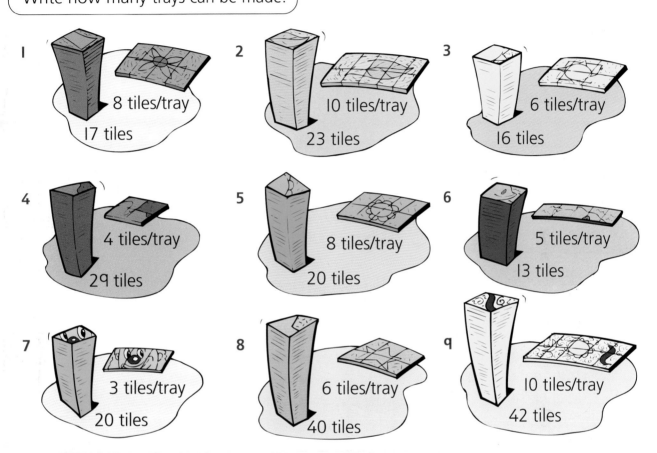

1 8 tiles/tray — 17 tiles

2 10 tiles/tray — 23 tiles

3 6 tiles/tray — 16 tiles

4 4 tiles/tray — 29 tiles

5 8 tiles/tray — 20 tiles

6 5 tiles/tray — 13 tiles

7 3 tiles/tray — 20 tiles

8 6 tiles/tray — 40 tiles

9 10 tiles/tray — 42 tiles

ℯ How many more trays are needed to make 10 trays each time?

Explore

Use number cards 1 to 9.

Use three cards to make a mixed number, e.g. $2\frac{1}{4}$.

Write the mixed number as an improper fraction.

How many different mixed numbers and improper fractions between 2 and 4 can you make?

Write all the mixed numbers in order.

Copy and complete.

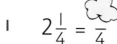

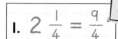

1. $2\frac{1}{4} = \frac{9}{4}$

1   $2\frac{1}{4} = \frac{\square}{4}$

2   $2\frac{4}{7} = \frac{\square}{7}$

3   $3\frac{2}{9} = \frac{\square}{9}$

4   $\square\frac{2}{5} = \frac{17}{5}$

5   $3\frac{6}{11} = \frac{\square}{11}$

6   $\square\frac{1}{2} = \frac{9}{2}$

7   $3\frac{3}{4} = \frac{\square}{8}$

8   $1\frac{4}{5} = \frac{\square}{10}$

9   $4\frac{2}{3} = \frac{\square}{6}$

10 $1\frac{5}{7} = \frac{\square}{14}$

11 $2\frac{1}{6} = \frac{\square}{12}$

12 $6\frac{1}{8} = \frac{\square}{16}$

13 Each wooden puzzle has **16** identical pieces.
Luka has **40** pieces.
How many puzzles does he have?

Problems

14 A flask of juice holds **4** cups.
Sukki has **5** full flasks and one which is $\frac{3}{4}$ full.
How many cups can she fill?

15 Rolls of ribbon are cut into **eighths** for decorations.
Katie has $2\frac{1}{2}$ rolls of ribbon.
How many decorations can she make?

16 Logs are chopped into **fifths** for firewood.
Tim has **18** pieces of firewood.
How many whole logs has he chopped?

Tenths, hundredths, thousandths

Write the number of tenths, hundredths and thousandths.

1 0·386

1. 3 tenths
8 hundredths
6 thousandths

2 1·725

3 2·868

4 3·079

5 1·908

6 7·357

7 4·15

8 6·002

9 3·308

10 9·167

11 4·039

12 2·776

13 1·307

Write the number of metres.

14 3 dm and 4 mm

14. 0·3 0 4 m

15 2 dm, 4 cm and 6 mm

16 5 dm, 6 cm and 1 mm

17 3 cm and 9 mm

18 4 dm, 3 cm and 8 mm

19 2 dm, 2 cm and 2 mm

20 7 cm and 2 mm

21 8 cm and 9 mm

22 6 cm and 8 mm

Write the value of the 9 in each quantity.

1. $9\,ml = \dfrac{9}{1000}\,l$

1  0·409 l

2  0·219 l

3  0·493 l

4  0·039 l

5 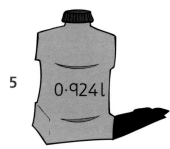 0·924 l

6  0·393 l

e Write each capacity in millilitres and centilitres.

Write the number of kilograms.

7. 0·0 0 8 kg

7  8 g

8  62 g

9  20 g

10  300 g

11  500 g

12  670 g

13  31 g

14  87 g

15  4 g

Ordering decimals

Write each set of distances in order, smallest to largest.

1
1·305 m
1·35 m
1·503 m

1. 1·3 0 5 m, 1·3 5 m, 1·5 0 3 m

2
0·62 m
0·602 m
0·629 m

3
0·901 m
0·009 m
0·01 m

4
4·04 m
4·004 m
4·104 m

5
0·303 m
0·33 m
0·033 m

6
1·001 m
1·101 m
1·01 m

7
2·34 m
2·43 m
2·403 m

8
0·171 m
0·071 m
0·17 m

9
4·64 m
4·4 m
4·406 m

10
5·92 m
5·092 m
5·29 m

Explore

Find how many 3-place decimal numbers there are between 1·5 and 1·6.

How many of these contain a 9?

1·5 < 1·501 < 1·6
1·5 <

Mixed problems

1 Amazing 99s

Complete these:

$15 \times 99 =$ 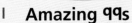 $16 \times 99 =$

$17 \times 99 =$ $18 \times 99 =$

Use the pattern to write the answers to 19×99, 20×99.

Does this work for any number?

2 Davinder goes on holiday for **2 weeks.**

He has **£500** to spend.

He pays **£18** per night to stay in a guesthouse. How much is the total bill?

He spends **£112** on food. Approximately how much spending money does he have for each day of his holiday?

3 Sally is training for a long-distance race. The race is 12 laps of the track.

Sally runs **1** lap on the first day, **2** laps on the second day, **3** laps on the third day, … until she can run **12** laps on the twelfth day. After this Sally runs **12** laps each day.

How many laps does she run in total in her first **4** weeks of training?

4

Try this five times.
What do you notice?

Think of a number under 100. → Multiply it by 9.

Look at your answer.

Halve it FOUR times! ← Double it.

Take away double the number you started with.

Subtract 16.

5 Multiples of 5
Try this trick.

$\boxed{2}\boxed{5} \times \boxed{2}\boxed{5} = \boxed{6}\boxed{2}\boxed{5}$

add 1

$\boxed{2} \times \boxed{3} = \boxed{6}$ $\boxed{5} \times \boxed{5} = \boxed{2}\boxed{5}$

Try this for 35×35:

$3 \times 4 = 12$ $5 \times 5 = 25$

$35 \times 35 = 1225$

Use the pattern to write the answers to 45×45, 55×55 and 65×65.

6 Matthew's dog Poppy needs to walk between **5 km** and **8 km** each day to keep fit.

What is the fewest number of kilometres Matthew could walk Poppy in June? What is the largest number?

Matthew walks Poppy **6·2 km** every day for the first fortnight in June. How much further must he walk her so Poppy has walked the minimum distance for June?

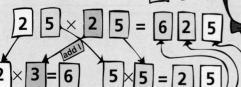